An Intelligence in Our Image

The Risks of Bias and Errors in Artificial Intelligence

Osonde Osoba, William Welser IV

For more information on this publication, visit www.rand.org/t/RR1744

Library of Congress Cataloging-in-Publication Data is available for this publication.
ISBN: 978-0-8330-9763-7

Published by the RAND Corporation, Santa Monica, Calif.
© Copyright 2017 RAND Corporation
RAND® is a registered trademark.

Cover: the-lightwriter/iStock/Getty Images Plus

Support RAND
Make a tax-deductible charitable contribution at
www.rand.org/giving/contribute

www.rand.org

Preface

Algorithms and artificial intelligence agents (or, jointly, *artificial agents*) influence many aspects of life: the news articles read, access to credit, and capital investment, among others. Because of their efficiency and speed, algorithms make decisions and take actions on the behalf of humans in these and many other domains. Despite these gains, there are concerns about the rapid automation of jobs (even cognitive jobs, such as journalism and radiology). This trend shows no signs of abating.

As reliance on artificial agents continues to grow, what are the consequences and risk of such dependence? A better understanding of attitudes toward and interactions with algorithms is essential precisely because of the aura of objectivity and infallibility today's culture ascribes to algorithms. This report illustrates some of the shortcomings of algorithmic decisionmaking, identifies key themes around the problem of algorithmic errors and bias (e.g., data diet, algorithmic disparate impact), and examines some approaches for combating these problems.

This report should be of interest to decisionmakers and implementers looking for a better understanding of how artificial intelligence deployment can affect their stakeholders. This affects such domains as criminal justice, public works, and welfare administration.

RAND Ventures

The RAND Corporation is a research organization that develops solutions to public policy challenges to help make communities throughout the world safer and more secure, healthier and more prosperous. RAND is nonprofit, nonpartisan, and committed to the public interest.

RAND Ventures is a vehicle for investing in policy solutions. Philanthropic contributions support our ability to take the long view, tackle tough and often-controversial topics, and share our findings in innovative and compelling ways. RAND's research findings and recommendations are based on data and evidence and, therefore, do not necessarily reflect the policy preferences or interests of its clients, donors, or supporters.

Funding for this venture was provided by the generous contributions of the RAND Center for Global Risk and Security (CGRS) Advisory Board. The research was conducted within CGRS, part of International Programs at the RAND Corporation.

Support for this project is also provided, in part, by the fees earned on client-funded research.

RAND Center for Global Risk and Security

CGRS draws on RAND's extensive multidisciplinary expertise to assess the impact of political, social, economic, and technological trends on global risk and security.

For more information about the RAND Center for Global Risk and Security, visit www.rand.org/international/cgrs or contact the center director (contact information is provided on the center's web page).

Contents

Figure

Acknowledgments

We would like to thank Andrew Parasiliti for sponsoring this work through the RAND Center for Global Risk and Security. We also want to thank fellow researchers who provided pointed advice as this work evolved, particularly Bart Kosko, Angela O'Mahony, Sarah Nowak, and John Davis. Finally, we would like to thank our reviewers, John Seely Brown and Timothy R. Gulden, for their insightful feedback.

CHAPTER ONE

Introduction

Algorithms and artificial intelligence (AI) agents (or, jointly, *artificial agents*) influence many aspects of our lives: the news articles we read, the movies we watch, the people we spend time with, our access to credit, and even the investment of our capital. We have empowered them to make decisions and take actions on our behalf in these and many other domains because of the efficiency and speed gains they afford. Despite these benefits, we hear concerns about the rapid automation of jobs (even cognitive jobs, such as journalism and radiology). This automation trend shows no signs of abating.[1]

As our reliance on artificial agents continues to grow, so does the risk. A better understanding of our attitudes and interactions with algorithms is essential precisely because of the aura of objectivity and infallibility our culture ascribes to algorithms (Bogost, 2015). What happens when we let what Goddard, Roudsari, and Wyatt (2012) called *automation bias* run rampant? Algorithmic errors in commercial recreational systems may have limited impact (e.g., sending someone on a bad date). But errant algorithms in infrastructure (electrical grids), defense systems, or financial markets could contain potentially high global security risk. The "Flash Crash" of 2010 illustrates how vulnerable our reliance on artificial agents can make us (Nuti et al., 2011). The White House's Office of Science and Technology Policy has noted the threat our increasing reliance on opaque artificial agents presents

---

[1] Autor (2010) highlighted a 12.6-percent decline in middle-skill automation-sensitive jobs from 1979–2009. Jaimovich and Siu (2012) found that these middle-class job losses accelerate in recessions and do not recover.

to privacy, civil rights, and individual autonomy, warning about the "potential of encoding discrimination in automated decisions" (Executive Office of the President, 2016, p. 45) and discussing the problem features and case studies in such areas as credit reporting, employment opportunities, education, and criminal justice (see also Executive Office of the President, 2014). It is important to evaluate the extent and severity of that threat.

Our goal here is to explain the risk associated with uncritical reliance on algorithms, especially when they implicitly or explicitly mediate access to services and opportunities (e.g., financial services, credit, housing, employment). Algorithmic decisions are not automatically equitable just by virtue of being the products of complex processes, and the procedural consistency of algorithms is not equivalent to objectivity. DeDeo (2015, p. 1) describes this issue succinctly: "[algorithms] may be mathematically optimal but ethically problematic." While human decisionmaking is also rife with comparable biases that artificial agents might exhibit, the question of accountability is murkier when artificial agents are involved.

The rest of this report takes the following structure. Chapter Two defines and examines the concept of an algorithm. Then we turn our attention to complex algorithms behaving incorrectly or inequitably. Our primary focus will be on the impact of artificial agents in social and policy domains. Chapter Three steps away from particular examples to dissect the issues underlying the problem of misbehaving algorithms. We will propose a selection of remedies to reclaim a measure of accountability for algorithmic decisionmaking processes. This includes recent work on fair, accountable, and transparent machine learning. In Chapter Four, we conclude with some observations and recommendations on how to better understand and address the challenges of algorithmic bias.

CHAPTER TWO

Algorithms: Definition and Evaluation

People are often unclear on the nature of the algorithms controlling large portions of their lives. Decisionmakers and policy analysts increasingly rely on algorithms as they try to make timely effective decisions in a data-rich world. Their use of algorithms (or artificial agents more generally) as decision aids encapsulates details that are important but not pertinent to the decision. This is a strong benefit of algorithmic aids for decisionmaking. A properly functioning algorithm frees up the decisionmaker's cognitive capacity for other important deliberations.[1]

But the opacity of algorithms makes it harder to judge correctness, evaluate risk, and assess fairness in social applications. It can also obscure the causal understanding behind decisions. These issues might be harmless if algorithms were (near) infallible. But most algorithms have only probabilistic guarantees of accuracy. And this is in the best possible scenarios, in which the right models and algorithms are applied appropriately, with the best intention to "perfect" data. Algorithm designers and users rarely have the luxury of such perfect scenarios. They must rely on assumptions that can fail and lead to unexpected results.[2]

---

[1] Procedural consistency is an argument for this aided decisionmaking model. The use of algorithms limits the effect of subjective or arbitrary decisionmaking. But Citron (2007, p. 1252) argued that the extensive use of automation and algorithmic decision aids has led to digital systems being "the primary decision-makers in public policy" instead of decision aids in some areas of administrative law. She also raises related questions about due process: Decisions made algorithmically may offer limited avenues for legitimate appeal or redress.

[2] For example, Salmon (2012) argued that the 2008 financial crash was a result of overreliance on an inaccurate model of default risk correlation, the Gaussian copula.

The fallibility of algorithms is an easy point to make. This includes systematic algorithmic errors, not just the statistical inaccuracies inherent to many algorithms. There are many examples in public policy–oriented applications. As a concrete example of significant error, Google's Flu Trends tool is famous for repeatedly misdiagnosing nationwide flu trends (Lazer et al., 2014). Many risk-estimation algorithms were based on incorrect probabilistic models and failed to react properly just before the 2008 U.S. financial crash (Salmon, 2012). One city implemented algorithms intended to optimally detect street potholes based on passively collected data from smartphone users. The demographic breakdown of smartphone users at the time would have led to blind spots, causing some communities to be underserved (Crawford, 2013). This would have had the effect of depriving less-affluent citizens' access to city repair services. Another city decided to use algorithmic approaches to direct its law-enforcement activities. The justification was that predictive policing algorithms were more objective as they only relied on objective "multi-variable equations," not on subjective human decisions (quoted in Tett, 2014). Reporting on another criminal justice application, Angwin et al. (2016) demonstrated systematic bias in a criminal risk assessment algorithm used in sentencing hearings across the United States.

Defining Algorithms

It will be helpful to carefully examine what algorithms are as we proceed. The concept has shifted quite a bit over centuries. The medieval Islamic scholar Abu-Abdullah Muhammed ibn-Musa Al-Khwarizmi, who lent his name to the algorithm was more interested in reliable step-by-step procedures for computing solutions to equations (Arndt, 1983). Alonzo Church and Alan Turing (Turing, 1937a; Turing, 1937b) introduced the concepts of *computability* and *computable functions* to formalize the idea of an algorithm. The definition amounted to a finite sequence of precise instructions that are implementable on computing systems (including but not limited to human brains). This probably brings to mind involved rote procedures, such as recipes for making a dish or steps for calculating your federal tax burden. Church and Turing's definitions lead directly to the common conception of algorithms as just code for crunching numbers.

The late Marvin Minsky (1961) and other pioneering AI thinkers (such as John McCarthy and Frank Rosenblatt) whose work followed Church and Turing were thinking about a different aspect of algorithms: empowering computing systems with the gift of intelligence. A prime hallmark of intelligence is the ability to adapt or learn inductively from "experience" (i.e., data). Their efforts led to the formulation of *learning* algorithms for training computing systems to learn and/or create useful internal models of the world. These algorithms also consist of rote sequential computational procedures at the microscopic level. The difference is that the algorithms are not just crunching numbers through static mathematical models but update their behavior iteratively based on models tuned in response to their experience (input data) and performance metrics.[3] Yet the problem of learning remains notoriously hard.[4] Many of the initial algorithms tried to mimic biological behaviors.[5] The grand goal was (and still is) to create autonomous AI capable of using such advanced learning algorithms to rival or exceed fluid human intelligence. Such systems are often called *general AI* systems in current discussions. Commercial successes—such as Google's recent AlphaGo triumph (Silver et al., 2016) and Micro-

---

[3] Valiant (2013) makes the argument that evolution itself is a type of learning algorithm iteratively adapting biological and social traits to improve a reproductive fitness performance metric.

[4] The problem of learning to distinguish between truth and falsehood based on experience is more formally known as *the problem of induction*. The central question is how justifiable applying generalizations based on limited past experience to new scenarios is. Philosophers have given much thought to this problem. David Hume in particular expressed concerns about the use of induction for learning about causality (Hume, 2000, Sec. VII). Bertrand Russell explains the point with the example of a chicken who has learned to identify the farmer as the cause of its daily meals based inductively on extensive past experience. It has no reason to expect the farmer to be the cause of its final demise (Russell, 2001).

[5] There was an initial flurry of interaction between AI pioneers and psychologists (both behaviorist and physiologically inclined) to try to understand how animals learn new behaviors.

soft's advanced AI chatbot, Tay (Lee, 2016)—show how far this line of research has come.

Much of the AI pioneers' work forms the foundation of machine learning algorithms that underpin most of the automated systems used today. These automated systems typically focus on learning to solve "simpler" tasks, such as automatic speech and image recognition. The common term for such systems is *narrow AI.* Their success is partly attributable to the exponential explosion in computational power available for implementing and extending their algorithms. For example, their work forms the basis of state-of-the-art deep learning methods used for modern image and speech recognition.[6]

The ongoing "big data" revolution also serves as a powerful catalyst promoting the wide use of learning algorithms. Big data (Brown, Chui, and Manyika, 2011) provides the steady stream of multimodal data necessary for extracting valuable insight via learning algorithms. This stream will only grow as objects become more networked (e.g., in an "Internet of Things") to produce more data. The only sustainable way to make sense of the sheer volume and variety of data produced daily is to apply powerful algorithms.

Our cultural conception of algorithms tends to conflate the full spectrum of algorithms from blind computation procedures (i.e., static computations) to advanced automated learning and reasoning procedures used in such systems as IBM's Watson (Ziewitz, 2016). Ian Bogost (2015) argued that this cultural conception of algorithms is sloppy shorthand that encourages laypersons to treat algorithms as monolithic, opaque, almost theological constructs. Many of the key algorithms that affect public life are also considered proprietary or trade secrets. Veils of secrecy do not tend to promote well-informed public discourse.

This opaque, uninformed understanding of algorithms impedes intelligent public discourse on their shortcomings. For example, how do we discuss questions about the validity of algorithms given the large variety of them? On one hand, the validity of a blind computational

---

[6] *Deep learning* refers to the use of many hidden processing layers in connectionist (made up of connected processing elements) machine learning architectures (such as neural networks).

algorithm is a function of how correct its implementation is. For example, does an algorithm for calculating tips correctly implement percentage multiplication and addition? Does an algorithm for calculating a tax burden take proper account of taxable income and apply the right rules according to the tax code? Did the sorting algorithm actually sort the entire data set or ignore parts of it? These are questions concerning concrete, sometimes objectively verifiable concepts.

But the validity of a learning algorithm is a somewhat different creature. It is a function of the correctness of its implementation (what algorithm designers tend to focus on) and the correctness of its learned behavior (what lay users care about). As a recent example, take Microsoft's AI chatbot, Tay. The algorithms behind Tay were properly implemented and enabled it to converse in a compellingly human way with Twitter users. Extensive testing in controlled environments raised no flags. A key feature of its behavior was the ability to learn and respond to user's inclinations by ingesting user data. That feature enabled Twitter users to manipulate Tay's behavior, causing the chatbot to make a series of offensive statements (Lee, 2016). Neither its experience nor its data took novelty in a new context into account.

This type of vulnerability is not unique to this example. Learning algorithms tend to be vulnerable to characteristics of their training data. This is a feature of these algorithms: the ability to adapt in the face of changing input. But algorithmic adaptation in response input data also presents an attack vector for malicious users. This *data diet vulnerability* in learning algorithms is a recurring theme.

"Misbehaving" Algorithms: A Brief Review

As artificial agents take a larger role in decisionmaking processes, more attention needs to be paid to the effects of fallible and misbehaving artificial agents.

Artificial agents are, by definition, not human. Moral judgment typically requires an element of choice, empathy, or agency in the actor. There can be no meaningful morality associated with artificial agents;

their behavior is causally determined by human specification.[7] The term *misbehaving algorithm* is only a metaphor for referring to artificial agents whose results lead to incorrect, inequitable, or dangerous consequences.

The history of such misbehaving artificial agents extends at least as far back as the advent of ubiquitous computing systems. Batya Friedman and the philosopher Helen Nissenbaum (1996) discussed bias concerns in the use of computer systems for tasks as diverse as scheduling, employment matching, flight routing, and automated legal aid for immigration. Friedman and Nissenbaum's discussion was nominally about the use of computer systems. But their critique was aimed at the procedures these systems used to generate their results: algorithms. Friedman and Nissenbaum's analyses reported inequitable or biased behavior in these algorithms and proposed a systematic framework for thinking about such biases.

Friedman and Nissenbaum (1996) wrote about the Semi-Automated Business Reservations Environment (SABRE) flight booking system, which American Airlines had sponsored (see also Sandvig et al., 2014). SABRE provided an industry-changing service. It was one of the first algorithmic systems to provide flight listings and routing information for airline flights in the United States. But its default information sorting behavior took advantage of typical user behavior to create a systematic anticompetitive bias for its sponsor.[8] SABRE always presented agents with flights from American Airlines on the first page, even when other airlines had cheaper or more-direct flights for the same query. Nonpreferred flights were often relegated to the second and later pages, which agents rarely reached. American Airlines

---

[7] Many of the debates over liability in automated systems revolve around this question: *What degree of AI autonomy is sufficient to limit the ethical responsibility of human administrators for the consequences of the AI's actions?* For example, to what extent is a company, such as Google, Facebook, or Tesla, liable for unforeseeable second-, third-, or higher-order effects of using its automated systems? How do we delimit foreseeability for a system that is necessarily (at least for now) opaque? The Google defamation example we offer later shows law courts beginning to grapple with such questions on the limits of liability.

[8] American Airlines' employees took to calling the practice *screen science.*

was forced to make SABRE more transparent after antitrust proceedings shed light on these concerns.

Friedman and Nissenbaum (1996) also examined the history of the algorithm for the National Resident Match Program, which matches medical residents to hospitals throughout the United States. The algorithm's seemingly equitable assignment rules favored hospital preferences over resident preferences and single residents over married residents.[9] Friedman and Nissenbaum also looked at the British Nationality Act Program, which was designed to encode British citizenship law. The act itself has fairness issues. And any faithful algorithm implementing the act has inherited, and thus magnified, these issues. The more interesting point was that the British Nationality Act Program presented authoritative responses that hid relevant legal options in the act from nonexpert users. The program's responses were procedurally correct. But translating the law into an exact algorithm lost important nuances.

The systems Friedman and Nissenbaum reported on were the larger, industrial-scale systems common in the early days of personal computing and the internet. The exponential growth of the internet and the personal computer user base expanded the scope of these problems. Algorithms began to mediate more of our interactions with information. Google is the canonical case in point. Google's search and advertising placement algorithms were digesting massive amounts of user-generated data to learn to optimize service for users (both regular users and advertising entities). Such systems were some of the first to expose the results of learning algorithms to widespread personal consumption.

---

[9] This example refers to the National Resident Match Program matching algorithm in use before Alvin Roth changed it in the mid-90s (Roth, 1996). This matching procedure was an incarnation of the stable matching algorithm first formalized by David Gale and Lloyd Shapley (Gale and Shapley, 1962). It is stable in the sense that neither hospital nor resident can do better, given the whole groups' stated preference order. However, contrary to initial claims, the program led to *stable* matches that guaranteed the hospitals their best acceptable choices but only guaranteed acceptable choices for students (sometimes their least acceptable ones).

Latanya Sweeney and Nick Diakopoulos pioneered the study of misbehavior in Google systems (Sweeney, 2013; Diakopoulos, 2013; Diakopoulos, 2016). Their work exposed instances of *algorithmic defamation* in Google searches and ads. Diakopoulos discussed a canonical example of such algorithmic defamation in which search engine autocompletion routines, fed a steady diet of historical user queries, learn to make incorrect defamatory or bigoted associations about people or groups of people.[10] Sweeney showed that such learned negative associations affect Google's targeted ads. In her example, just searching for certain types of names led to advertising for criminal justice services, such as bail bonds or criminal record checking. Diakopoulos's examples included consistently defamatory associations for searches related to transgender issues.

Studies like Sweeney's and Diakopoulos's are archetypes in the growing field of data and algorithmic journalism. More news and research articles chronicle the many missteps of the algorithms that affect different parts of our lives, online and off. IBM's *Jeopardy*-winning AI, Watson, famously had to have its excessive swearing habit corrected after its learning algorithms ingested some unsavory data (Madrigal, 2013). There have also been reports on the effects of Waze's traffic routing algorithms on urban traffic patterns (Bradley, 2015). One revealing book describes the quirks of the data and algorithms underlying the popular OkCupid dating service (Rudder, 2014). More recently, former Facebook contractors revealed that Facebook's news feed trend algorithm was actually the result of subjective input from a human panel (Tufekci, 2016).

Others began writing on the effects of algorithms in governance, public policy, and messy social issues. Artificial agents have to contend with another layer of complexity and peril in these spaces. Bad behavior here could have far-reaching, populationwide, and generational consequences.

Citron (2007) reported on how the spread of algorithmic decisionmaking to legal domains deprives citizens of due process. More

---

[10] Germany now holds Google partially responsible for the correctness of its autocomplete suggestions (Diakopoulos, 2013).

recently, Angwin et al. (2016) reported on extreme systematic bias in a criminal risk assessment algorithm in widespread use in sentencing hearings across the country.

Citron and Pasquale (2014) wrote about what they call *the scored society* and its pitfalls. By *scored society*, they mean the current state in which unregulated, opaque, and sometimes hidden algorithms produce authoritative scores of individual reputation that mediate access to opportunity. These scores include credit, criminal, and employability scores. Citron and Pasquale particularly focused on how such systems violate reasonable expectations of due process, especially expectations of fairness, accuracy, and the existence of avenues for redress. They argue that algorithmic credit scoring has not reduced bias and discriminatory practices. On the contrary, such scores serve to legitimize bias already existing in the data (and software) used to train the algorithms. Pasquale (2015) followed a similar line of inquiry with an exhaustive report on the extensive use of unregulated algorithms in three specific areas: reputation management (e.g., credit scoring), search engine behavior, and the financial markets.

Barocas and Selbst (2016) wrote a recent influential article addressing the fundamental question of whether big data can result in fair or neutral behavior in algorithms.[11] They argue that the answer to this question is a firm negative without reforming how big data and the associated algorithms are applied. Barocas and Selbst (and other researchers in this field) borrow *disparate impact* from legal doctrine originally introduced in the 1960s and 1970s to test the fairness of employment practices.[12] The authors use the term to denote the systematic disadvantages artificial agents impose on subgroups based on patterns learned via procedures that appear reasonable and nondiscriminatory on face value. Gandy (2010) used *rational discrimination* to refer to an identical concept. He was arguing for the need of regu-

---

[11] Barocas and Selbst use *big data* as an umbrella term for the massive data sets and algorithmic techniques used to analyze such data.

[12] In *Griggs* v. *Duke Power Co.* (1971), the U.S. Supreme Court ruled against Duke Power Company's practice of using certain types of employment tests and requirements that were unrelated to the job. Such tests may be innocuous on face value. But, on closer inspection, these tests "operate invidiously" to discriminate on the basis of race.

latory constraints on decision support systems to address the runaway negative externalities (such as cumulative disadvantages) these systems foster.

Barocas and Nissenbaum (2014) discussed how algorithms and big data also circumvent any legal privacy guarantees we have grown to expect. The standard safeguard against algorithmic disparate impact effects is to hide sensitive data fields (such as gender and race) from learning algorithms. But the literature on modern reidentification techniques recognizes that learning algorithms can implicitly reconstruct sensitive fields and use these probabilistically inferred proxy variables for discriminatory classification (DeDeo, 2015; Feldman et al., 2015). The power of these inference techniques only grows as more data sets are added to the training base (Ohm, 2010). This poses a problem for regulation; it is possible to legislate against the explicit use of protected information (such as race and gender in the Equal Employment Opportunity and Fair Housing acts).[13] But it is harder to legislate against the use of probabilistically inferred information. Pasquale (2015) reported that data fusion agencies already take advantage of this regulatory loophole.

Algorithm designers and researchers have recently begun to work on technical approaches to certify and/or remove algorithmic disparate impacts. Feldman et al. (2015) presented an approach to certifying that a classification algorithm is fair according to U.S. legal standards. Their correction procedure performs rank-preserving modifications to the input data to control disparate impact. DeDeo (2015) presented a method for modifying the output of an algorithm to decorrelate its output from protected variables. Dwork et al. (2012) leveraged some of Dwork's own insights on privacy (Dwork, 2008a; Dwork, 2008b) to develop a theoretical framework for fair classification algorithms. This approach looks for context-sensitive fair similarity metrics for comparing and classifying individuals regardless of protected category membership.

---

[13] The Fair Housing Act a set forth in Title VIII of the Civil Rights Act of 1968 and is set forth in 42 U.S. Code 3504–3606.

A Case Study: Artificial Agents in the Criminal Justice System

The U.S. criminal justice system is increasingly resorting to algorithmic tools. Artificial agents help ease the burden of managing such a large system. But any systematic algorithmic bias in these tools would have a high risk of errors and cumulative disadvantage.

We first look at the use of algorithms at the sentencing and parole phase. Angwin et al. (2016) reported on Northpointe's Correctional Offender Management Profiling for Alternative Sanctions (COMPAS) criminal risk assessment system. This software is used in sentencing and parole hearings across the country. Angwin et al. presented anecdotes showing the system misrepresenting the recidivism risks of different convicts. These anecdotes first hint at a systematic racial bias in the risk estimation. Black convicts were being rated higher than nonblack convicts, even when the nonblack convicts had more-severe offenses. The authors follow up on this hint with analysis of COMPAS and recidivism data from Broward County, Florida.

Larson et al. (2016), detailing the statistical analysis in Angwin et al. (2016), found that

> Black defendants were twice as likely as white defendants to be misclassified as a higher risk of violent recidivism, and white recidivists were misclassified as low risk 63.2% more often than black defendants.

Police departments are also resorting to algorithmic tools for predictive policing and allocating resources. We present an example of a simulation showing how a mathematically acceptable algorithm results in inequitable predictive policing behavior. Figure 1 illustrates how a mathematically effective algorithm for finding criminals based on historical crime data can lead to inequitable behavior. The figure illustrates the simulated behavior of an automated system for directing law enforcement efforts in finding and responding to criminal events in the whole population.

Suppose we have a population that splits naturally along categories (e.g., location, gender, crime type, or any other criterion) and that

Figure 1
Rate of Enforcement Events per Epoch: Two Subpopulations, Same Crime Rate, Differing Vigilance

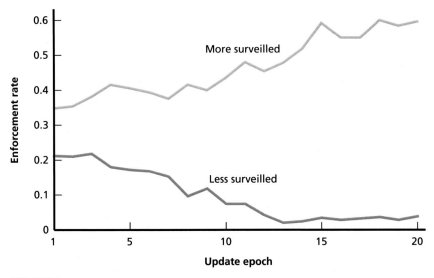

RAND *RR1744-1*

we have limited enforcement resources that cannot find and respond to all criminal events. Thus, for a population with a given crime rate, we find violations and enforce the law with a given probability. A vigilance parameter determines how much attention we pay in the different classes and thus the probability of successfully finding and prosecuting crimes when they do occur. The system's vigilance adapts using a mathematically reasonable learning procedure. The system increases vigilance in areas with higher perceived propensity for criminality and/ or in areas with higher enforcement activity based on recorded enforcement data.[14]

The algorithm is reasonably effective but easily leads to inequitable outcomes. The plot in the figure shows a simulation of its behavior giving an increasingly disproportionate frequency of enforcement events in the distinct subpopulations. Such behavior would be accept-

---

[14] The learning law for the algorithm is similar to the ideas behind the *broken windows* approach to law enforcement advocated by Wilson and Kelling (1982).

able if it held only when underlying levels of criminality were also disproportionate. But the figure shows that this subpopulation-specific divergent behavior can still hold when the underlying subpopulation-specific crime rates are the same. This has the effect of increasingly criminalizing specific subpopulations and (more critically) generating more "objective" data to support future biased enforcement decisions.

The algorithm described is more than just an illustrative tool. It also simulates the effect of inequitable and unprincipled state surveillance based on historic records. The simulation makes the argument that increased state surveillance is not a neutral tool, especially if it is not uniformly applied. On a populationwide-level, it can lead to *inequitable criminalization*, in which criminals with different demographic characteristics have systematically different likelihoods of apprehension and different sentencing intensities. The recent "color of surveillance" movement makes the argument that state surveillance in the United States has not been equitably applied. Some legal scholars argue that inequitable criminalization is the norm in the United States, often justified based on historical records of crime, as the COMPAS system was. The U.S. Department of Justice (2016) found more evidence of such inequitable surveillance and criminalization in its recent investigation of the Baltimore Police Department.

A similar algorithm could be applied to the problem of finding fruitful mining or oil-drilling sites. Such an algorithm's "inequitable" behavior—giving more attention to areas with a historical record of producing oil—may count as a feature (more focused attention and resources applied productively) and not a bug in that context. Part of the difference is that solutions to questions of public policy often need to account for other measures of quality informed by (sometimes fuzzy or inexact) social principles, such as equity or fairness. In this case, we expect law enforcement to be fair in the sense that enforcement activity should be proportional to criminal activity across categories.

CHAPTER THREE

The Problem in Focus: Factors and Remedies

The examples in the previous chapters illustrate a number of angles on the algorithmic bias problem. The first and most basic angle is the problem of an algorithm's data diet: With limited human direction, an artificial agent is only as good as the data it learns from. Automated learning on inherently biased data leads to biased results. The agent's algorithms try to extract patterns from data with limited human input during the act of extraction. The limited human direction makes a case for the objectivity of the process. But data generation is often a social phenomenon (e.g., social media interactions, online political discourse) inflected with human biases. Applying procedurally correct algorithms to biased data is a good way to teach artificial agents to imitate whatever bias the data contains. For example, recent research shows that automated methods applied to language necessarily learn human biases inherent in our use of language (Caliskan-Islam, Bryson, and Narayanan, 2016).

This leads to the rather paradoxical effect that artificial agents, learning autonomously from human-derived data, will often learn human biases—both good and bad. We could call this *the paradox of artificial agency.* The Watson and Tay examples illustrate the point well. Sweeney (2013) also gives multiple examples of targeted advertising systems making biased and sometimes defamatory inferences about particular individuals because of biases automatically learned from data. This paradox has important implications for the use of artificial agents in the big data era. The complexity of data patterns and the sheer scale of available data make it necessary for artificial agents to learn more autonomously. This suggests that individuals should expect more artificial agents to mirror human biases.

The second angle of the algorithmic bias problem often applies when working with policy or social questions. This is the difficulty of defining ground truth or identifying robust guiding principles. Our ground truth or criteria for judging correctness are often culturally or socially informed, as the IBM Watson and Google autocomplete examples illustrate. Learning algorithms would need to optimize over some measure of social acceptability in addition to whatever performance metrics they are optimizing internally to perform a task. This dual optimization easily leads to dilemmas. In fact, the recent work on fair algorithms shows that there is usually a trade-off between accuracy and fairness. Enforcing fairness constraints can mean actively occluding or smearing informative variables. This can reduce the strength of algorithmic inference.

Another angle on the problem is that judgments in the space of social behavior are often fuzzy, rather than well-defined binary criteria.[1] This angle elaborates on the second point. The examples presented earlier show fuzzy cultural norms ("do not swear," "do not bear false witness," "present a balanced perspective") influencing human judgment of correct algorithmic behavior. We are able to learn to navigate complex fuzzy relationships, such as governments and laws, often relying on subjective evaluations to do this. Systems that rely on quantified reasoning (such as most artificial agents) can mimic the effect but often require careful design to do so. Capturing this nuance may require more than just computer and data scientists.

Another system has evolved over centuries to answer policy questions subject to fuzzy social norms and conflicting reports or data: the law. Grimmelmann and Narayanan (2016) pointed out that, while

---

[1] Here, *fuzzy* has a precise meaning, referring to properties and sets that have inexact boundaries of definition. It is based on the idea of multivalued (i.e., not binary) logic and set membership pioneered by such thinkers as Max Black (in his work on vague sets) and Lotfi Zadeh (in his work on fuzzy logic). As a concrete example, think about the set of temperatures you would call "warm." The border between the temperature sets "warm" and "not-warm" is inexact. In the swearing AI examples we discuss, swearing is neither absolutely forbidden nor absolutely permissible; its social acceptability exists on a spectrum.

crypto currencies and algorithmic ("smart") contracts might excel at enforcing binary property rights, property rights in the real world are fuzzy and contentious. Similar concerns apply to algorithms: What we consider proper algorithmic behavior can sometimes be only defined imprecisely. The law has evolved for adjudicating such fuzzy complexities.

U.S. law also recognizes that procedures that are reasonable on face value can have adverse and disparate impact. An understanding of this concept of disparate impact is only slowly spreading in the algorithm research community. There is an increasing body of work on the social and legal impact of data and algorithms (Gangadharan, Eubanks, and Barocas, 2015). And a growing body of evidence shows that algorithms do not automatically treat diverse populations fairly and equitably just by virtue of being reasonable algorithms (Barocas and Selbst, 2016; DeDeo, 2015; Dwork et al., 2012; Feldman et al., 2015; Hardt, 2014).

Other Technical Factors

Other technical factors besides those we have already discussed promote algorithmic bias. Machine learning algorithms have issues handling sample-size disparities. This is a direct consequence of the fact that machine learning algorithms are inherently statistical methods and are therefore subject to the statistical sample-size laws. Learning algorithms may have difficulty capturing specific cultural effects when the population is strongly segmented. This is related to the problem of statistical inference on highly nonstationary training data (particularly when default models do not account for nonstationary effects).

Sample-Size Disparity

Machine learning algorithms are statistical estimation methods. Their measures of estimation error often vary in inverse proportion with data sample sizes. This means that these methods will typically be more error-prone on low-representation training classes than with others. A credit-estimation algorithm would be more error-prone on subpopulations that have historically low representation in credit markets. Dwoskin (2015) reported on a concrete demonstration of this effect. Yahoo's automated image tagging system made racist image-labeling choices precisely because of demographic inhomogeneity in its training data. For a good discussion of sample-size disparity, see Hardt, 2014.

Hacked Reward Functions

Reward functions in machine learning and AI theory come from behaviorist psychology, as in the work of B. F. Skinner. These functions are the principal means by which current artificial learning systems learn correct behavior. During an artificial agent's learning process, the reward function quantifies how much we reward or punish good or bad actions and decisions. Learning algorithms then adapt the agent's parameters and behavior to maximize total reward. Thus, the design of AI behavior often reduces to the design of sufficiently incentivizing reward functions. This behaviorist approach to learning can be gamed. For example, a cleaning robot designed to minimize the amount of dirt it sees may gain rewards for just shutting down its visual sensors instead of cleaning. Amodei et al. (2016) refers to this process as *reward hacking*. A poorly specified reward function can lead to undesirable side effects or behaviors in AI systems. Reward hacking is also a concern as humans adapt their behaviors to algorithmic evaluation. People learn to game algorithms given enough exposure (e.g., learning which cheap, irrelevant signals credit-scoring systems factor into a credit score).

Cultural Differences

Machine learning algorithms work by selecting salient features (variables) in the data that telegraph or correlate with various behaviors (Hardt, 2014). Behaviors that are culturally mediated may lead to inequitable behavior. Hardt gives the example of how cultural differences in naming conventions led to flagging of accounts with nontraditional names on social media platforms.[2]

---

2 This is a cultural phenomenon often called *Nymwars*. Such platforms as Twitter, Google+, and Blizzard Entertainment (game developer) have argued that having real names attached to accounts helps maintain decorum online. So, they have actively flagged and/or deleted

Confounding Covariates

Algorithm designers often choose to remove sensitive variables from their training data in an attempt to render the resulting system bias free (Barocas and Selbst, 2016). A common refrain by system designers is that "the system cannot be biased because it does not take [some sensitive variable] into account." Barocas and Selbst (2016) discuss why just hiding sensitive variables often does not solve the problem. Machine learning methods can often probabilistically infer hidden variables, e.g., using ZIP Codes to infer (as a probabilistic substitute for) income. This ability also has strong implications for data privacy and anonymity. Researchers are reporting that traditional expectations of data privacy and anonymity are no longer attainable (Dwork, 2008a; Narayanan and Shmatikov, 2010; Ohm, 2010). This is because modern machine learning algorithms are capable of reidentifying data easily and robustly.

Remedies

Remedying or regulating artificial agents will most likely require a combination of technical and nontechnical approaches. There are recent efforts underway to develop fair, accountable, and transparent machine learning techniques. We propose augmenting these with less technical approaches.

Statistical and Algorithmic Approaches

There is a growing field focused on fair, accountable, and transparent machine learning, working on technical approaches to assuring algorithmic fairness or certifying and correcting disparate impact in machine learning algorithms. Dwork et al. (2012) proposed using modified distance or similarity metrics when working with subject data. These similarity metrics are meant to enforce rigorous fairness constraints when comparing subjects in data sets. Sandvig et al. (2014) proposed a number of algorithms auditing procedures that

---

accounts with seemingly false names. The process for distinguishing between real and false names relied heavily on traditional Western naming practices (Hardt, 2014; Boyd, 2012).

compare algorithmic output with expected equitable behavior. Algorithm audits can be more feasible and thorough when algorithm codes and procedures are open sourced. DeDeo (2015) introduced an algorithmic approach to ensuring that machine learning models enforce statistical independence between outcomes and protected variables. Feldman et al. (2015) introduced a test for checking whether an algorithm violates legal disparate impact rules (under U.S. law). This provides a socially informed metric of optimality. They also proposed a statistical method for correcting such inequities in classification algorithms. Yet, there is a drawback: These schemes will often trade some predictive power for fairness.

Causal Reasoning Algorithms

More broadly and on a longer time-scale, Judea Pearl (2009), Leon Bottou et al. (2013), and others (Athey, 2015) are exploring ways to equip machine learning algorithms with causal or counterfactual reasoning. This is extremely important because automated causal reasoning systems can present clear causal narratives for judging the quality of an algorithmic decision process. Accurate causal justifications for algorithmic decisions are the most reliable audit trails for algorithms.

A precedent-setting U.S. Supreme Court case on capital punishment illustrates the importance of causal reasoning in decisionmaking (*McCleskey* v. *Kemp*, 1987). A legal scholar, David Baldus, had explored the use of quantitative empirical methods to test the excessiveness of capital sentencing decisions in California (Baldus et al., 1980). Baldus then applied his analysis to the state of Georgia in his 1983 study (Baldus, Pulaski, and Woodworth, 1983). The study used carefully controlled statistical analyses of observational data on capital punishment to illustrate the disproportionate impact of capital sentencing for the state of Georgia.[3] Baldus's exhaustive analysis included about 230 variables.

---

[3] Baldus, Pulaski, and Woodworth (1983) found, in summary, that race increased the odds (by a multiplicative factor of 4.3) of a capital sentence for otherwise comparable convictions in the state of Georgia.

The court case involved dueling statistical experts debating the findings of Baldus's study. The court proceedings even included protracted discussions on detailed statistical concepts, such as multicollinearity.

The Supreme Court finally held that the sentencing was valid because the study did not demonstrate deliberate bias in McCleskey's case. This was the final decision in spite of the carefully demonstrated 4-to-1 racial disparity in sentencing outcomes. The court's justification was that, however true the Baldus study was, it did not demonstrate that race was a *causal factor* in McCleskey's particular sentencing.

If we are to rely on algorithms for autonomous decisionmaking, they need to be equipped with tools for auditing the causal factors behind key decisions. Algorithms that can be audited for causal factors can give clearer accounts or justifications for their outcomes. This is especially important for justifying statistically disproportionate outcomes.

Algorithmic Literacy and Transparency

Combating algorithmic bias would benefit from an educated public capable of understanding that algorithms can lead to inequitable outcomes. This is not the same as requiring that users understand the inner workings of all algorithms—this is not feasible. Just instilling a healthy dose of informed skepticism could be useful enough to reduce the effect size of automation bias. There is hope on this front. The sheer amount of time we spend interfacing with algorithms may make algorithmic missteps more noticeable.

For example, online dating users (a rapidly rising percentage of the population) routinely question the results of date matching algorithms. Journalism and documentaries on the 2008 subprime mortgage financial crash have also helped foster a healthy, more-informed cultural skepticism about the efficacy of complex algorithms. Consider recent reports of public outcry over the SketchFactor app (Marantz, 2015). The app used crowdsourced data and aggregation algorithms to calculate a neighborhood's "sketchiness" score. There was significant negative reaction to the app on the ground of cultural insensitivity and potential for discriminatory abuse. The public was able to clearly articulate concerns that the algorithmic SketchFactor score would likely encode preexisting cultural biases about neighborhoods.

Combining algorithmic literacy with transparency could be very effective. Transparency in this space usually refers to making sure any algorithms in use are easily understood. Again, that is unlikely to be feasible all the time. What is feasible and useful is more disclosure of decisions and actions that are mediated by artificial agents. Users seem to already expect such disclosures as the backlash over Facebook's new curation practices indicates (Tufekci, 2016). Beyond just satisfying user tastes, disclosure-style transparency allows users to be better informed and more skeptical consumers of information.

Personnel Approaches

The technical research on bias in machine learning and artificial intelligence algorithms is still in its infancy. Questions of bias and systemic errors in algorithms demand a different kind of wisdom from algorithm designers and data scientists. These practitioners are often engineers and scientists with less exposure to social or public policy questions. The demographics of algorithm designers are often less than diverse. Algorithm designers make myriad design choices, some of which may have far-reaching consequences. Diversity in the ranks of algorithm developers could help improve sensitivity to potential disparate impact problems.

However the drive to remedy algorithmic bias should be tempered with a healthy dose of regulatory restraint. Any kind of remedy would require algorithms to adhere more closely to socially defined values. Which values, and who gets to decide? Questions about free speech, censorship, equitability, and other acceptable ethical standards will need to be addressed as society wades deeper into these waters.

CHAPTER FOUR

Conclusion

We have illustrated the challenge of algorithmic disparate impact, why we can expect to see an expansion of algorithm dependence, and what the best options for mitigating future risk might be. The error and bias risk in algorithms and AI will continue as long as artificial agents play increasingly prominent roles in our lives and remain unregulated.

Response to unregulated artificial agents tends to be of three broad types: avoiding algorithms altogether, making the underlying algorithms transparent, or auditing the output of algorithms. Avoiding algorithms is probably impossible; few other options are available for making sense of the current deluge of data. Algorithmic transparency requires a more educated public capable of understanding algorithms. But recent advances in deep connectionist learning mean that, even if we could deconstruct an algorithm's procedure, it may still be too complex to make useful sense of that insight.

Christian Sandvig's recent work argues that the last option, the algorithm audit, should be the way forward (Sandvig et al., 2014). Certain audit types ignore the inner workings of artificial agents and judge them according to the fairness of their results. This is akin to how we often judge human agents: by the consequences of their outputs (decisions and actions) and not on the content or ingenuity of their code base (thoughts). This option makes the most sense for policymakers and sets the standard for a consequentialist ethics for artificial agents. Regulation is much easier under this framing.

Discussions like this one may sometimes anthropomorphize artificial agents: Are machines beginning to think like us, and how can we judge and guide them? Current progress on artificial agents may make this anthropomorphic view of algorithms closer to the norm. This may have the unexpected benefit of fostering public understanding that artificial agents, like humans, are not above bias.

Abbreviations

AI   artificial intelligence

COMPAS   Correctional Offender Management Profiling for Alternative Sanctions

IBM   International Business Machines, Inc.

SABRE   Semi-Automated Business Reservations Environment

References

Amodei, Dario, Chris Olah, Jacob Steinhardt, Paul Christiano, John Schulman, and Dan Mané, "Concrete Problems in AI Safety," Ithaca, N.Y.: Cornell University Library, 2016. As of February 2, 2017: https://arxiv.org/abs/1606.06565

Angwin, Julia, Jeff Larson, Surya Mattu, and Lauren Kirchner, "Machine Bias: There's Software Used Across the Country to Predict Future Criminals. And It's Biased Against Blacks," *ProPublica*, May 23, 2016. As of December 5, 2016: https://www.propublica.org/article/machine-bias-risk-assessments-in-criminal-sentencing

Arndt, A. B., "Al-Khwarizmi," *Mathematics Teacher*, Vol. 76, No. 9, 1983, pp. 668–670.

Athey, Susan, "Machine Learning and Causal Inference for Policy Evaluation," *Proceedings of the 21st ACM SIGKDD International Conference on Knowledge Discovery and Data Mining*, Sydney, Australia, August 10–13, 2015, pp. 5–6.

Autor, David, "The Polarization of Job Opportunities in the U.S. Labor Market: Implications for Employment and Earnings," Washington, D.C.: Center for American Progress and the Hamilton Project, April 2010.

Baldus, David C., Charles Pulaski, and George Woodworth, "Comparative Review of Death Sentences: An Empirical Study of the Georgia Experience," *Journal of Criminal Law and Criminology*, Vol. 74, No. 3, Autumn 1983, pp. 661–753.

Baldus, David C., Charles A. Pulaski, George Woodworth, and Frederick D. Kyle, "Identifying Comparatively Excessive Sentences of Death: A Quantitative Approach," *Stanford Law Review*, Vol. 33, No. 1, November 1980, pp. 1–74.

Barocas, Solon, and Helen Nissenbaum, "Big Data's End Run Around Procedural Privacy Protections," *Communications of the ACM,* Vol. 57, No. 11, 2014, pp. 31–33.

Barocas, Solon, and Andrew D. Selbst, "Big Data's Disparate Impact," *California Law Review*, Vol. 104, 2016, pp. 671–732.

Bogost, Ian, "The Cathedral of Computation," *Atlantic*, January 15, 2015. As of May 27, 2016:
http://www.theatlantic.com/technology/archive/2015/01/the-cathedral-of-computation/384300/

Bottou, Léon, Jonas Peters, Joaquin Quiñonero-Candela, Denis X. Charles, D. Max Chickering, Elon Portugaly, Dipankar Ray, Patrice Simard, and Ed Snelson, "Counterfactual Reasoning and Learning Systems: The Example of Computational Advertising," *Journal of Machine Learning Research*, Vol. 14, No. 1, 2013, pp. 3207–3260.

Boyd, Danah, "The Politics of Real Names," *Communications of the ACM*, Vol. 55, No. 8, 2012, pp. 29–31.

Bradley, Ryan, "Waze and the Traffic Panopticon," *New Yorker*, June 2, 2015. As of December 2, 2016:
http://www.newyorker.com/business/currency/waze-and-the-traffic-panopticon

Brown, Brad, Michael Chui, and James Manyika, "Are You Ready for the Era of 'Big Data'?" *McKinsey Quarterly*, Vol. 4, No. 1, October 2011, pp. 24–35.

Caliskan-Islam, Aylin, Joanna J. Bryson, and Arvind Narayanan, "Semantics Derived Automatically from Language Corpora Necessarily Contain Human Biases," Ithaca, N.Y.: Cornell University Library, 2016. As of February 2, 2017:
https://arxiv.org/abs/1608.07187

Citron, Danielle Keats, "Technological Due Process," *Washington University Law Review*, Vol. 85, No. 6, 2007, pp. 1249–1313.

Citron, Danielle Keats, and Frank A. Pasquale, "The Scored Society: Due Process for Automated Predictions," *Washington Law Review*, Vol. 89, 2014.

Crawford, Kate, "Think Again: Big Data," *Foreign Policy*, Vol. 9, May 10, 2013.

DeDeo, Simon, "Wrong Side of the Tracks: Big Data and Protected Categories," Ithaca, N.Y.: Cornell University Library, May 28, 2015. As of March 7, 2017:
https://arxiv.org/pdf/1412.4643v2.pdf

Diakopoulos, Nicholas, "Algorithmic Defamation: The Case of the Shameless Autocomplete," Tow Center for Digital Journalism website, August 6, 2013. As of February 14, 2017:
http://towcenter.org/algorithmic-defamation-the-case-of-the-shameless-autocomplete/

———, "Accountability in Algorithmic Decision Making," *Communications of the ACM*, Vol. 59, No. 2, 2016, pp. 56–62.

Dwork, Cynthia, "An ad Omnia Approach to Defining and Achieving Private Data Analysis," in Francesco Bonchi, Eleena Ferrari, Bradley Malin, and Yücek Saygin, eds., *Privacy, Security, and Trust in KDD*, New York: Springer, 2008a, pp. 1–13.

————, "Differential Privacy: A Survey of Results," *International Conference on Theory and Applications of Models of Computation*, New York: Springer, 2008b, pp. 1–19.

Dwork, Cynthia, Moritz Hardt, Toniann Pitassi, Omer Reingold, and Richard Zemel, "Fairness Through Awareness," *Proceedings of the 3rd Innovations in Theoretical Computer Science Conference*, Cambridge, Mass., January 8–10, 2012, pp. 214–226.

Dwoskin, Elizabeth, "Social Bias Creeps into New Web Technology," *Wall Street Journal*, August 21, 2015. As of December 5, 2016:
http://www.wsj.com/articles/
computers-are-showing-their-biases-and-tech-firms-are-concerned-1440102894

Equal Employment Opportunity Act of 1972. As of March 8, 2017:
https://www.eeoc.gov/eeoc/history/35th/thelaw/eeo_1972.html

Executive Office of the President, *Big Data: Seizing Opportunities, Preserving Values*, Washington, D.C., May 2014. As of February 3, 2017:
https://obamawhitehouse.archives.gov/sites/default/files/docs/big_data_privacy_report_5.1.14_final_print.pdf

————, *Preparing for the Future of Artificial Intelligence*, Washington, D.C., October 2016. As of February 3, 2017:
https://obamawhitehouse.archives.gov/sites/default/files/whitehouse_files/microsites/ostp/NSTC/preparing_for_the_future_of_ai.pdf

Feldman, Michael, Sorelle A. Friedler, John Moeller, Carlos Scheidegger, and Suresh Venkatasubramanian, "Certifying and Removing Disparate Impact," *Proceedings of the 21st ACM SIGKDD International Conference on Knowledge Discovery and Data Mining*, Sydney, Australia, August 10–13, 2015, pp. 259–268.

Friedman, Batya, and Helen Nissenbaum, "Bias in Computer Systems," *ACM Transactions on Information Systems,* Vol. 14, No. 3, July 1996, pp. 330–347.

Gale, David, and Lloyd S. Shapley, "College Admissions and the Stability of Marriage," *American Mathematical Monthly*, Vol. 69, No. 1, January 1962, pp. 9–15.

Gandy, Oscar H., Jr., "Engaging Rational Discrimination: Exploring Reasons for Placing Regulatory Constraints on Decision Support Systems," *Ethics and Information Technology,* Vol. 12, No. 1, March 2010, pp. 29–42.

Gangadharan, Seeta Peña, V. Eubanks, and S. Barocas, *Data and Discrimination: Collected Essays*, Washington, D.C.: Open Technology Institute, 2015.

Goddard, Kate, Abdul Roudsari, and Jeremy C. Wyatt, "Automation Bias: A Systematic Review of Frequency, Effect Mediators, and Mitigators," *Journal of the American Medical Informatics Association,* Vol. 19, No. 1, January–February 2012, pp. 121–127.

Griggs v. *Duke Power Co.*, 401 U.S. 424, 1971.

Grimmelmann, James, and Arvind Narayanan, "The Blockchain Gang," *Slate*, February 16, 2016. As of December 5, 2016:
http://www.slate.com/articles/technology/future_tense/2016/02/bitcoin_s_blockchain_technology_won_t_change_everything.html

Hardt, Moritz, "How Big Data Is Unfair: Understanding Sources of Unfairness in Data Driven Decision Making," *Medium*, September 26, 2014. As of December 5, 2016:
https://medium.com/@mrtz/how-big-data-is-unfair-9aa544d739de

Hume, David, *An Enquiry Concerning Human Understanding: A Critical Edition*, Tom L. Beauchamp, ed., Vol. 3, New York: Oxford University Press, 2000.

Jaimovich, Nir, and Henry E. Siu, "The Trend Is the Cycle: Job Polarization and Jobless Recoveries," paper, Cambridge, Mass.: National Bureau of Economic Research, 2012.

Larson, Jeff, Surya Mattu, Lauren Kirchner, and Julia Angwin, "How We Analyzed the COMPAS Recidivism Algorithm," *ProPublica*, May 23, 2016. As of December 6, 2016:
https://www.propublica.org/article/how-we-analyzed-the-compas-recidivism-algorithm

Lazer, David, Ryan Kennedy, Gary King, and Alessandro Vespignani, "Google Flu Trends Still Appears Sick: An Evaluation of the 2013–2014 Flu Season," paper, Rochester, N.Y.: Social Science Electronic Publishing, Inc., March 13, 2014. As of February 2, 2017:
https://papers.ssrn.com/sol3/papers.cfm?abstract_id=2408560

Lee, Peter, "Learning from Tay's introduction," blog, Microsoft website, March 25, 2016. As of December 5, 2016:
http://blogs.microsoft.com/blog/2016/03/25/learning-tays-introduction/#sm.0001v8vtz3qddejwq702cv2annzcz

Madrigal, Alexis C., "IBM's Watson Memorized the Entire 'Urban Dictionary,' Then His Overlords Had to Delete It," *Atlantic*, January 10, 2013. As of December 5, 2016:
http://www.theatlantic.com/technology/archive/2013/01/ibms-watson-memorized-the-entire-urban-dictionary-then-his-overlords-had-to-delete-it/267047/

Marantz, Andrew, "When an App Is Called Racist," *New Yorker*, July 29, 2015. As of December 5, 2016:
http://www.newyorker.com/business/currency/what-to-do-when-your-app-is-racist

McCleskey v. *Kemp*, 481 U.S. 279, 1987.

Minsky, Marvin, "Steps Toward Artificial Intelligence," *Proceedings of the IRE*, Vol. 49, No. 1, 1961, pp. 8–30.

Narayanan, Arvind, and Vitaly Shmatikov, "Myths and Fallacies of Personally Identifiable Information," *Communications of the ACM,* Vol. 53, No. 6, June 2010, pp. 24–26.

Nuti, Giuseppe, Mahnoosh Mirghaemi, Philip Treleaven, and Chaiyakorn Yingsaeree, "Algorithmic Trading," *Computer,* Vol. 44, No. 11, 2011, pp. 61–69.

Ohm, Paul, "Broken Promises of Privacy: Responding to the Surprising Failure of Anonymization," *UCLA Law Review,* Vol. 57, 2010, p. 1701.

Pasquale, Frank, *The Black Box Society: The Secret Algorithms That Control Money and Information,* Boston: Harvard University Press, 2015.

Pearl, Judea, *Causality,* Cambridge: Cambridge University Press, 2009.

Roth, Alvin E., "The NRMP as a Labor Market: Understanding the Current Study of the Match," *Journal of the American Medical Association,* Vol. 275, No. 13, 1996, pp. 1054–1056.

Rudder, Christian, *Dataclysm: Love, Sex, Race, and Identity—What Our Online Lives Tell Us About Our Offline Selves,* New York: Crown Publishing, 2014.

Russell, Bertrand, *The Problems of Philosophy,* New York: Oxford University Press, [1912] 2001.

Salmon, Felix, "The Formula That Killed Wall Street," *Significance,* Vol. 9, No. 1, February 2012, pp. 16–20.

Sandvig, Christian, Kevin Hamilton, Karrie Karahalios, and Cedric Langbort, "Auditing Algorithms: Research Methods for Detecting Discrimination on Internet Platforms," paper presented to the Data and Discrimination: Converting Critical Concerns into Productive Inquiry preconference of the 64th Annual Meeting of the International Communication Association, Seattle, Wash., May 22, 2014.

Silver, David, Aja Huang, Chris J. Maddison, Arthur Guez, Laurent Sifre, George Van Den Driessche, Julian Schrittwieser, Ioannis Antonoglou, Veda Panneershelvam, and Marc Lanctot, "Mastering the Game of Go with Deep Neural Networks and Tree Search," *Nature,* Vol. 529, No. 7587, January 28, 2016, pp. 484–489.

Sweeney, Latanya, "Discrimination in Online Ad Delivery," *ACM Queue,* Vol. 11, No. 3, April 2, 2013, p. 10.

Tett, Gillian, "Mapping Crime—Or Stirring Hate?" *Financial Times,* August 22, 2014. As of December 6, 2016:
https://www.ft.com/content/200bebee-28b9-11e4-8bda-00144feabdc0

Tufekci, Zeynep, "The Real Bias Built in at Facebook," *New York Times,* May 19, 2016. As of December 6, 2016:
http://www.nytimes.com/2016/05/19/opinion/the-real-bias-built-in-at-facebook.html

Turing, Alan M., "Computability and λ-Definability," *Journal of Symbolic Logic*, Vol. 2, No. 4, 1937a, pp. 153–163.

———, "On Computable Numbers, with an Application to the Entscheidungsproblem," *Proceedings of the London Mathematical Society*, Vol. 2, No. 1, 1937b, pp. 230–265.

42 U.S. Code 3504, Discrimination in the Sale or Rental of Housing and Other Prohibited Practices, 1968.

42 U.S. Code 3505, Discrimination in Residential Real Estate–Related Transactions, 1968.

42 U.S. Code 3506, Discrimination in the Provision of Brokerage Services, 1968.

U.S. Department of Justice, Civil Rights Division, *Investigation of the Baltimore City Police Department*, Washington, D.C., August 10, 2016. As of December 6, 2016:
https://www.justice.gov/opa/file/883366/download

Valiant, Leslie, *Probably Approximately Correct: Nature's Algorithms for Learning and Prospering in a Complex World*, New York: Basic Books, 2013.

Wilson, James Q., and George L. Kelling, "Broken Windows: The Police and Neighborhood Safety," *Atlantic Monthly*, Vol. 249, No. 3, March 1982, pp. 29–38.

Ziewitz, Malte, "Governing Algorithms: Myth, Mess, and Methods," *Science, Technology, & Human Values*, Vol. 41, No. 1, September 30, 2016, pp. 3–16.